THE LITTLE FUN BOOK of

SPIDERS/NEANDERTHAL

BY

John Hodgson

This book is a work of fiction. Places, events, and situations in this story are purely fictional. Any resemblance to actual persons, living or dead, is coincidental.

ISBN: 1-4033-8756-7 (e-book)
ISBN: 1-4033-8757-5 (Paperback)

Library of Congress Control Number: 2002095295

This book is printed on acid free paper.

Printed in the United States of America
Bloomington, IN

1stBooks – rev. 10/30/02

PREFACE

This book is intended to look at the relationship

between two entities and how they interact in certain

situations.

A spider/Neanderthal spins a web/builds

a home.

A spider/Neanderthal devours its prey by

catching it in a web/spears and

sometimes ropes.

Spiders/Neanderthals hide to catch their

prey/bush.

When spiders/Neanderthals hunt they
pretend they're not waiting patiently/in a
bush.

A spider/Neanderthal spins a web/wore

silk from web.

.

A spider/Neanderthal sits and waits/gets

bored.

Spiders/Neanderthal had no

newspapers/didn't have knowledge.

Spiders/Neanderthal hangs from

trees/hides in crevices.

Spiders/Neanderthal eat food/spiders

take what it can steal.

Spiders/Neanderthal builds fire from

wood/hide behind wood.

Spiders/Neanderthal will sleep under

shelter/does not like when it rains.

John Hodgson

Spiders/Neanderthal plays with

stones/will bite a bee.

Spiders/Neanderthal have bad teeth/goes

through much grief.

Spiders/Neanderthal don't like each

other/are recluse.

Spiders/Neanderthal live in caves/always

fight.

Spiders/Neanderthal jump when they see

fire/lightening strikes.

Spiders/Neanderthals can not count/has

eight legs.

Spiders/Neanderthals can't speak.

Spiders/Neanderthal run into

cave/scurries when scared.

Spiders/Neanderthal cry when hungry/eat flies.

Spiders/Neanderthal get in trouble/get

eaten.

John Hodgson

Spiders/Neanderthals get bored/always

bored.

Neanderthals sometimes eat spiders.

Spiders sometimes bite Neanderthal.

Spiders/Neanderthal can learn from one
another.

Spiders/Neanderthals are very

jumpy/clumsy.

Spiders/Neanderthals do what they need

to/likes what it does.

Spiders/Neanderthals think each other

are funny.

Spiders/Neanderthals live in the same

place /cave.

Spiders/Neanderthal find each other to be

very different.

Neanderthals did not learn all of silk's

secrets.

Spider/Neanderthal formed a

relationship/both live in cave.

Spiders/Neanderthals were the first

geniuses put on earth/happiness.

Spiders bite only when threatened or

injured.

Spiders/Neanderthals are the largest

group of arachnids/many packs.

Spiders/Neanderthals are not yet fully

identified/misunderstood.

Spiders/Neanderthals are predators/hunt.

Spiders/Neanderthals don't all spin

webs/produce silk.

Spiders/Neanderthals were free/modern

day man.

Spiders/Neanderthal were born with

special skills/hunting.

Spiders/Neanderthal survived well on

earth/fruit and vegetables.

Spiders/Neanderthal sometimes found

food/potato in ground.

Spiders/ Neanderthals learned how to

prepare food/fire.

Spiders/Neanderthals have a lot of

knowledge as they had time to

think/nothing else to do.

John Hodgson

Spiders/Neanderthal found fire on

earth/lightening striking wood.

Spiders/Neanderthal still marvel us

today/disappeared.

There are still mysteries to be found

about spiders/Neanderthal.

Spiders/Neanderthals were

coexistent/sleeping in the same cave.

Spiders/Neanderthal produced

offspring/staying close.

Spiders/Neanderthal were close/giving

each other pain.

Spiders/Neanderthals lived and

hunted/predators.

Spiders/Neanderthal had no

inventions/only each other.

Spiders/Neanderthals did not have any

light at night/city lights.

Spiders/Neanderthals leaned on each

other for comfort/friends.

Spiders/Neanderthal took care of each

other/cave.

Spiders/Neanderthal bite into their

prey/fangs.

Spiders/Neanderthal took pride in what

they did/enjoyment.

Spiders/Neanderthal slept deep/nothing

to bother them.

John Hodgson

Spiders/Neanderthal had war between

each other/biting.

Spiders/Neanderthal had no eating

utensils/rocks.

Spiders/Neanderthal watched one

another closely/children.

Spiders/Neanderthal have had faults of

their own/did not try hard enough.

Spiders/Neanderthals had personal

problems/grief.

Spiders/Neanderthals should have joined

forces/learned from one another.

John Hodgson

Spiders/Neanderthals shared the same

habitat/cave.

Neanderthal got bite by spider in

cave/made weapon.

Spider was responsible for Neanderthals

extinction/cave.

Spider/Neanderthal lived together/didn't

always get along.

Spiders/Neanderthals ate many kinds of apples/lived in apples.

Spiders/Neanderthal did what they

thought was right/lived together.

Spiders/Neanderthal did not have

schools/they were what they were.

Spiders/Neanderthal sometimes cried at

night/tooth ache.

Spiders/Neanderthal cannot

communicate with the same

language/they understood each other in

cave.

Spiders/Neanderthals took their day

off/slept.

Spider was troubled by

Neanderthal/sleeping habits/cave.

Neanderthal did not see many she

Neanderthals/ no television/no travel.

John Hodgson

Spiders/Neanderthals time on earth

wasn't wasted/building blocks of modern

man.

Spiders/Neanderthal were together for

along time/cave.

Spiders/Neanderthals played games with

each other/ate.

Spiders/Neanderthal focused on what

they could/paid attention.

John Hodgson

Spiders/Neanderthals formed

relationships/seen eye to eye.

Spiders/Neanderthal learned how to hide

from their predators/woolly mammoth.

Spiders/Neanderthal didn't have music to

listen to/video.

Spiders/Neanderthals environment

wasn't good/animals.

Spiders/Neanderthal didn't have dental

coverage/tooth decay.

Spiders/Neanderthal didn't know what

went on around the world/no newspaper,

no television.

Spiders/Neanderthal lived in the same

dwelling/cave.

Spiders/Neanderthals had laws in their

dwelling/group.

Neanderthals had leaders in the

families/parents.

Neanderthals were the building blocks of

modern man. Spider still exists.

John Hodgson

Spider/Neanderthal could not travel far

across the earth/no car.

Spiders/Neanderthal knew a lot about

one another/friends.

Spiders/Neanderthal both had comfort

with each other/in cave.

Spiders/Neanderthal didn't particularly

like each other/spent too much tine

together/cave.

Spider/Neanderthal left each other.

Neanderthal developed into modern man.

Spider still exists.

Spider/Neanderthal laughed with each

other/saying good-bye.

Spiders/Neanderthal came to conclusion about each other. Neanderthals left cave.

Neanderthals and spiders left each other

alone/modern man.

Spiders now wait to catch food/hide.

Spider/Neanderthal had good life

together/not much to do now.

John Hodgson

Spiders/Neanderthals didn't like each

other after all/extinct.

Spiders/Neanderthals relationship was

good/room for improvement.

John Hodgson

THE END

About the Author

He enjoys spending time learning new concepts using knowledge he has learned through studying relationships and interaction of human behavior in society and workplace.